# How to be Brilliant at Mental Arithmetic

Beryl Webber
Jean Haigh

Brilliant
PUBLICATIONS

We hope you and your class enjoy using this book. Other books in the series include:

If you would like further information on these or other titles published by Brilliant Publications, please write to the address given below or visit our website, www.brilliantpublications.co.uk.

Published by Brilliant Publications,
Unit 10, Sparrow Hall Farm, Edlesborough,
Bedfordshire LU6 2ES

email:      info@brilliantpublications.co.uk
website:    www.brilliantpublications.co.uk
general enquiries:
tel:        01525 222292

Written by Beryl Webber and Jean Haigh
Illustrated by Kate Ford

© Beryl Webber and Jean Haigh 1997
Printed ISBN 978 1 897675 21 2
ebook ISBN 978 0 85747 046 1

First published in 1997. Reprinted 1998, 2000, 2011.
10 9 8 7 6 5 4

# Contents

# Introduction

*How to be Brilliant at Mental Arithmetic* contains 41 photocopiable sheets designed to support the development of accurate mental recall of number facts and the skill to calculate mentally. They can be used whenever the need arises for particular activities to support and supplement whatever core mathematics programme you use. The activities provide learning experiences which can be tailored to meet individual children's needs.

The activities are addressed directly to the children. They are self-contained and many children will be able to work with very little support from you. You may have some children, however, who have the necessary mathematical concepts and skills but require your help in reading the sheets.

The children will need pencils and should be encouraged to do as much working as possible in their heads. Some activities require extra resources such as coloured pencils, counters, scissors and a calculator. Some will require the use of additional resource sheets and these can be found at the back of the book. Where this is the case, it has been indicated by a small box, with the relevant page number in it, in the top right hand corner, eg 47 .

*How to be Brilliant at Mental Arithmetic* relates directly to the programmes of study for Using and Applying Mathematics and Number. The page opposite gives further details, and on the contents page the activities are coded according to programme of study and difficulty. The difficulty is indicated by a letter code (A–C) and is provided to give you an indication of how the activities relate to mathematical progression within the key stage. Activities coded A are the most challenging. Each activity also relates to the Using and Applying section of the programme of study in a variety of ways.

Page 40 provides a self-assessment sheet so that children can keep a record of their own progress.

# Why teach mental arithmetic?

Mental arithmetic is an important aspect of mathematics. It has a central position in the National Curriculum programmes of study and children are now expected to be very adept at handling numbers mentally. The Key Stage 2 National Curriculum mathematics tests now include a separate test of mental arithmetic.

The skill of being able to calculate mentally quickly and accurately depends on a firm understanding of the four operations and sufficient grasp of the number system to allow the adoption of a variety of flexible techniques. Children need encouragement and active teaching to become good at discovering quick and efficient ways of calculating mentally. Everyone needs to develop a range of strategies with which they feel comfortable. When asked, a group of adults is likely to report that they undertake the same mental calculation in a wide range of ways. Each adopts those strategies that feel the most comfortable and familiar. In order to develop the range of strategies and for them to become comfortable and familiar, it is important that there are plenty of opportunities to practise the skills and extend the range available.

It might be thought that in the modern technological age where calculators are freely available the need for mental arithmetic would diminish. In fact, the reverse is true and the demand from employers for school leavers who are numerically agile, confident and accurate with mental calculations is growing. What seems to be happening is that the importance of written calculations is diminishing.

This book addresses the twin pillars of mental arithmetic, ie mental recall and mental agility. Mental recall depends on familiarity with number bonds and plenty of opportunity to practise. Mental agility depends more on confidence with the number system and good mental models of it. Mental calculation is *not* most easily done by trying to visualise it on what Hilary Shuard called 'the blackboard of the mind'. Children should be actively discouraged from doing this as it seriously inhibits quick and accurate mental arithmetic.

It is to help children develop good mental models of the number system that some activities in this book require the use of tools that may not usually be associated with mental arithmetic – such as 100 squares, base 10 apparatus and number lines. These tools provide alternative models for the number system that will help children develop a wider range of mental models. Calculators are also used in some activities for checking the accuracy of the mental calculations. They should not be used as a substitute for mental agility.

In this book there are basically three types of activity: those which help children develop a range of mental models; those which provide children with a range of strategies for mental calculation; and those which provide opportunity for practice of mental arithmetic. They all support the development of accurate mental recall of number facts and flexible and confident mental agility.

# Links to the National Curriculum

**The activities in this book allow children to have opportunities to:**

*   use and apply mathematics in practical tasks, in real-life problems and within mathematics itself;

*   take increasing responsibility for organizing and extending tasks;

*   develop flexible and effective methods of computation, and use them with understanding.

In particular these activities relate to the following sections of the Key Stage 2 programme of study.

**Using and Applying Mathematics**

**2      Making and monitoring decisions to solve problems**

     a      select and use the appropriate mathematics;

     b      try different mathematical approaches;

     c      develop their own mathematical strategies and look for ways to overcome difficulties;

     d      check their results and consider whether they are reasonable.

**3      Developing mathematical language and forms of communication**

     a      understand the language of
*   number
*   relationships, including 'multiple of', 'factor of'.

**4      Developing mathematical reasoning**

     b      search for patterns in their results;

     c      explain their reasoning.

**Number**

**2      Developing an understanding of place value and extending the number system**

     a      read, write and order whole numbers, understanding that the position of a digit signifies its value; use their understanding of place value to develop methods of computation, to approximate numbers to the nearest 10 or 100, and to multiply and divide by powers of 10 when there are whole number answers.

**3    Understanding relationships between numbers and developing methods of computation**

    a    explore number sequences, explaining patterns and using simple relationships;

    c    consolidate knowledge of addition and subtraction facts to 20; know the multiplication facts to 10 x 10; develop a range of mental methods for finding quickly from known facts those that they cannot recall; use some properties of numbers, including multiples, factors and squares;

    d    develop a variety of mental methods of computation to develop a range of non-calculator methods of computation that involve addition and subtraction of whole numbers, progressing to methods for multiplication and division of up to three-digit by two-digit whole numbers;

    f    understand and use the relationships between the four operations, including inverses.

**4    Solving numerical problems**

    a    develop their use of the four operations to solve problems, including those involving money and measures;

    b    choose sequences of methods of computation appropriate to a problem, adapt them and apply them accurately.

**Shape, Space and Measures**

**4    Understanding and using measures**

    a    know the rough metric equivalents of imperial units still in daily use.

---

Each activity has been coded on the contents page to indicate its main relationship with the above aspects of the programme of study for Key Stage 2.

The number and lower case letter represent the relevant sub-section and aspect of the programme of study.

For example:

N3(d) indicates Number, sub-section 3 (Understanding relationships between numbers and developing methods of computation), d - 'develop a variety of mental methods ...'

Each activity is also coded by an upper case letter (A-C) indicating the relative difficulty of the activity itself. Activities coded 'A' are the most challenging.

# Tens and units snap

**What you need**
A copy of the *Blank digit cards resource sheet*
(page 45), a calculator for checking.

Number the digit cards from 21 to 60 and cut them out.

Play the game with a friend.

Mix up the cards and deal them out equally between you.

Each turn over a card.

Both players add the two numbers and say the answer aloud. The first person to get the
right answer keeps the cards. You may need to use the calculator to check the answers.
Play continues until all the cards have been won. The winner is the person who has the most
cards at the end.

Record who wins here.

Remember to use all the methods
you have learned for working
things out in your head.

**EXTRA!**
Play the game using subtraction. Play the game using two pages
of digit cards numbered from 21 to 100.

---

How to be Brilliant at Mental Arithmetic

# Digit cards 0 – 9

### What you need
Two copies of the *Digit cards 0–9 resource sheet* (page 42), calculator.

Cut out the cards on the resource sheets.

Play with two friends. At each go two people play the game and the third person is the checker. Mix up the cards and lay them face down on the table.

The first player picks up two cards and, as quickly as possible, adds the two numbers together in his or her head and says the answer. The checker checks this on the calculator. If the answer is correct, the player ticks his or her first space in the record grid. If it is incorrect, the player puts a cross in the first place in the record grid. Play continues with the two players taking it in turns to turn over the cards and add the numbers until all the cards have been used up. The player with the most ticks at the end is the winner.

Have another go so that the checker can play this time.

| Turn | Player 1 | Player 2 |
|------|----------|----------|
| 1 | | |
| 2 | | |
| 3 | | |
| 4 | | |
| 5 | | |
| 6 | | |
| 7 | | |
| 8 | | |
| 9 | | |
| 10 | | |
| Total | | |

Play the game using subtraction. Then use multiplication.

**Tip:** Make sure when using subtraction you take the smaller number away from the larger.

### EXTRA!
Use a stop clock or watch to time each player answering the ten questions.
The player with the most correct in the faster time is the winner.

# Digit cards 0–19

**What you need**
A copy of the *Digit cards 0-19 resource sheet* (page 43), calculator.

Cut out the cards on the resource sheets.

Play with two friends. At each go two people play the game and the third person is the checker. Mix up the cards and lay them face down on the table.

The first player picks up two cards and, as quickly as possible, adds the two numbers together in his or her head and says the answer. The checker checks this on the calculator. If the answer is correct, the player ticks his or her first space in the record grid. If it is incorrect the player puts a cross in the first place in the record grid. Play continues with the two players taking it in turns to turn over the cards and add the numbers until all the cards have been used up. The player with the most ticks at the end is the winner.

Have another go so that the checker can play this time.

Play the game using subtraction.

**EXTRA!**
Use the stop clock or watch to time each player answering the ten questions. The player with the most correct in the faster time is the winner.

| Turn | Player 1 | Player 2 |
|---|---|---|
| 1 | | |
| 2 | | |
| 3 | | |
| 4 | | |
| 5 | | |
| 6 | | |
| 7 | | |
| 8 | | |
| 9 | | |
| 10 | | |
| 11 | | |
| 12 | | |
| 13 | | |
| 14 | | |
| 15 | | |
| 16 | | |
| 17 | | |
| 18 | | |
| 19 | | |
| 20 | | |
| Total | | |

How to be Brilliant at Mental Arithmetic

# Addition shortcuts, 1

This page will help you add numbers with units digits of 9 and 8 quickly in your head.

To work out the addition sum 67 + 9 you could add 10 and then subtract 1.

Remember:   10 = 9 + 1

So 67 + 9 ⟶ 67 + 10 − 1 ⟶ 76

Use this method to work out the following sums.

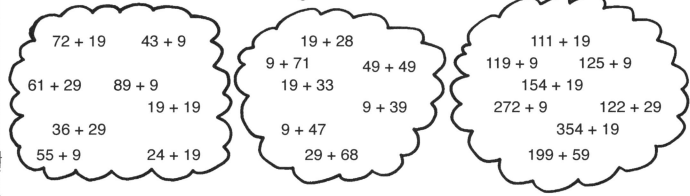

72 + 19      43 + 9

61 + 29      89 + 9

19 + 19

36 + 29

55 + 9      24 + 19

19 + 28

9 + 71      49 + 49

19 + 33

9 + 39

9 + 47

29 + 68

111 + 19

119 + 9      125 + 9

154 + 19

272 + 9      122 + 29

354 + 19

199 + 59

To work out the addition sum 67 + 8, you could add 10 and then subtract 2.

So 67 + 8 ⟶ 67 + 10 − 2 ⟶ 75

Use this method to work out the following sums.

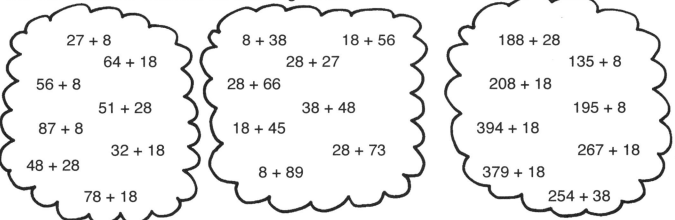

27 + 8

64 + 18

56 + 8

51 + 28

87 + 8

32 + 18

48 + 28

78 + 18

8 + 38      18 + 56

28 + 27

28 + 66

38 + 48

18 + 45

28 + 73

8 + 89

188 + 28

135 + 8

208 + 18

195 + 8

394 + 18

267 + 18

379 + 18

254 + 38

**EXTRA!**
Adapt this method to help you subtract numbers with units digits of 9 or 8 quickly.

# Addition shortcuts, 2

This page will help you add numbers with units digits of 6 and 7 quickly in your head.

To work out the addition sum 48 + 6 you could add 5 and then add 1.

Remember: 6 = 5 + 1

So 48 + 6 ——▶ 48 + 5 + 1 ——▶ 54

Use this method to work out the following sums.

29 + 96
21 + 6        64 + 26
73 + 76
59 + 6
82 + 36
37 + 16        19 + 36

86 + 15        26 + 11
56 + 27
6 + 19
16 + 39
6 + 91        36 + 72
96 + 42

123 + 16
245 + 36
842 + 96
107 + 6        121 + 16
197 + 86
352 + 26
429 + 56

To work out the addition sum 48 + 7 you could add 5 and then add 2.

So 48 + 7 ——▶ 48 + 5 + 2 ——▶ 55

Use this method to work out the following sums.

96 + 7
42 + 17        24 + 7
54 + 37
39 + 27
16 + 47
35 + 17
81 + 57

97 + 41        27 + 34
47 + 14
37 + 29
7 + 29
67 + 24
57 + 11
7 + 16

836 + 37
232 + 17
666 + 17
108 + 7
994 + 17
492 + 27
124 + 7
515 + 47

**EXTRA!**
Adapt this method to help you subtract numbers with units digits of 6 or 7 quickly.

# Product squares

**What you need**
A copy of the *Product square resource sheet* (page 44).

Look at the square below. What do you notice about the numbers?

| 1 | 2 | 3 |
|---|---|---|
| 2 | 4 | 6 |
| 3 | 6 | 9 |

Describe the patterns that go **across** the square.

Describe the patterns that go **down** the square.

Now complete another row and another column.

**Tip:** Try to describe the patterns as multiplication patterns.

| 1 | 2 | 3 | |
|---|---|---|---|
| 2 | 4 | 6 | |
| 3 | 6 | 9 | |
| | | | |

Complete the large product grid on the resource sheet.

Look for patterns within the square. Compare the patterns you have found with a friend. Try to find some different patterns.

Now look at this product square. What do you notice? Compare it with the one you did before. Now complete it.

| 10 | 20 | 30 | | |
|----|----|----|---|---|
| 20 | 400 | 600 | | |
| 30 | 600 | 900 | | |
| | | | | |
| | | | | |

**EXTRA!**
Use the large product square on the resource sheet and colour in all the odd numbers. What do you notice? Are there more odd or even numbers? Why?

How to be Brilliant at Mental Arithmetic

## What you need
Two copies of the *Quick recall resource sheet* (page 48), a calculator for checking.

**Tip:** Look as the last digits in the numbers in the sums. What can you say about them?

You will need to work with a friend.

One of you is the reader and the other answers the questions. The reader says the first sum in List A. The partner writes the answer in the first column of the resource sheet as quickly as possible. As soon as the first question has been answered, move on to the second and so on until all the questions in List A have been answered.

Then swap jobs and repeat, using List B.

When you have finished both lists, swap resource sheets and check each others' answers using a calculator.

You can have another go using Lists C and D.

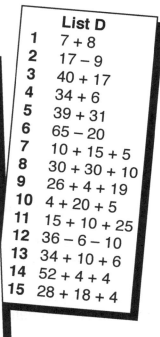

**List A**
1. 6 + 7
2. 15 − 9
3. 10 + 15
4. 23 + 7
5. 24 + 16
6. 42 − 20
7. 10 + 10 + 5
8. 20 + 5 + 5
9. 68 + 2 + 12
10. 2 + 10 + 5
11. 10 + 15 + 10
12. 39 − 9 − 10
13. 54 + 10 + 6
14. 62 + 4 + 4
15. 16 + 26 + 8

**List B**
1. 8 + 5
2. 13 − 7
3. 20 + 16
4. 46 + 4
5. 27 + 23
6. 59 − 30
7. 10 + 5 + 5
8. 20 + 20 + 5
9. 54 + 6 + 17
10. 3 + 20 − 6
11. 10 + 15 + 5
12. 48 − 8 − 20
13. 43 + 20 + 7
14. 34 + 3 + 3
15. 17 + 27 + 6

**List C**
1. 9 + 4
2. 14 − 8
3. 30 + 14
4. 58 + 2
5. 38 + 32
6. 76 − 40
7. 20 + 10 + 5
8. 30 + 20 + 5
9. 43 + 7 + 24
10. 3 + 20 + 6
11. 25 + 15 + 10
12. 47 − 7 − 20
13. 26 + 20 + 4
14. 44 + 3 + 3
15. 33 + 23 + 4

**List D**
1. 7 + 8
2. 17 − 9
3. 40 + 17
4. 34 + 6
5. 39 + 31
6. 65 − 20
7. 10 + 15 + 5
8. 30 + 30 + 10
9. 26 + 4 + 19
10. 4 + 20 + 5
11. 15 + 10 + 25
12. 36 − 6 − 10
13. 34 + 10 + 6
14. 52 + 4 + 4
15. 28 + 18 + 4

## EXTRA!
Make up some sums like those in the lists of your own. Each make up two lists of 15. Try to keep them as similar to those in the lists as possible. Look carefully at each sum to see the way it works.

Now try your own lists with your partner. Record your answers on the resource sheets as before.

# Quick recall, 2

**Tip:** Make sure your partner can understand the way you say the sums. For example, 8 x 7 can be said as 'eight times seven' or 'eight multiplied by seven' or 'seven eights'. Try to use 'multiplied by'.

### What you need
Two copies of the *Quick recall resource sheet* (page 48), a calculator for checking.

You will need to work with a friend.

One of you is the reader and the other answers the questions. The reader says the first sum in List A.
The partner writes the answer in the first column of the resource sheet as quickly as possible.
As soon as the first question has been answered, move on to the second and so on until all the questions in List A  have been answered.

Then swap jobs and repeat, using List B.

When you have finished both lists, swap resource sheets and check each others' answers using a calculator.

You can have another go using Lists C and D.

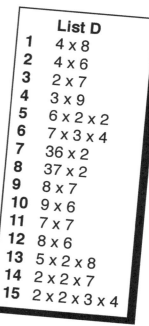

| | **List A** | | **List B** | | **List C** | | **List D** |
|---|---|---|---|---|---|---|---|
| 1 | 4 x 5 | 1 | 3 x 4 | 1 | 5 x 6 | 1 | 4 x 8 |
| 2 | 2 x 5 | 2 | 2 x 3 | 2 | 3 x 7 | 2 | 4 x 6 |
| 3 | 3 x 3 | 3 | 4 x 4 | 3 | 2 x 8 | 3 | 2 x 7 |
| 4 | 5 x 3 | 4 | 5 x 2 | 4 | 4 x 9 | 4 | 3 x 9 |
| 5 | 2 x 2 x 3 | 5 | 3 x 3 x 2 | 5 | 7 x 2 x 3 | 5 | 6 x 2 x 2 |
| 6 | 4 x 3 x 2 | 6 | 5 x 2 x 3 | 6 | 8 x 4 x 5 | 6 | 7 x 3 x 4 |
| 7 | 13 x 3 | 7 | 24 x 2 | 7 | 26 x 3 | 7 | 36 x 2 |
| 8 | 31 x 2 | 8 | 42 x 3 | 8 | 47 x 2 | 8 | 37 x 2 |
| 9 | 6 x 2 | 9 | 7 x 3 | 9 | 7 x 6 | 9 | 8 x 7 |
| 10 | 8 x 4 | 10 | 9 x 3 | 10 | 7 x 8 | 10 | 9 x 6 |
| 11 | 2 x 5 x 6 | 11 | 3 x 2 x 4 | 11 | 8 x 8 | 11 | 7 x 7 |
| 12 | 2 x 4 x 3 | 12 | 2 x 4 x 2 | 12 | 9 x 7 | 12 | 8 x 6 |
| 13 | 8 x 2 x 2 | 13 | 9 x 2 x 2 | 13 | 2 x 5 x 9 | 13 | 5 x 2 x 8 |
| 14 | 6 x 3 x 3 | 14 | 7 x 3 x 3 | 14 | 3 x 3 x 9 | 14 | 2 x 2 x 7 |
| 15 | 5 x 5 x 5 | 15 | 3 x 3 x 3 | 15 | 3 x 2 x 4 x 2 | 15 | 2 x 2 x 3 x 4 |

### EXTRA!
Make up some sums like those in the lists of your own. Each make up two lists of 15. Try to keep them as similar to those in the lists as possible. Look carefully at each sum to see the way it works.

Now try your own lists with your partner. Record your answers on the resource sheets as before.

How to be Brilliant at Mental Arithmetic

# Quick recall, 3

**Tip:** As you are working out each sum, try to see if there is a quick way you can work it out such as grouping numbers to make ten.

**What you need**
Two copies of the *Quick recall resource sheet* (page 48), a calculator for checking.

You will need to work with a friend.

One of you is the reader and the other answers the questions. The reader says the first sum in List A. The partner writes the answer in the first column of the resource sheet as quickly as possible. As soon as the first question has been answered move on to the second and so on until all the questions in List A have been answered.

Then swap jobs and repeat, using List B

When you have finished both lists, swap resource sheets and check each others' answers using a calculator.

You can have another go using Lists C and D.

**List A**

| 1 | 17 + 27 |
|---|---|
| 2 | 36 + 46 |
| 3 | 34 + 35 |
| 4 | 39 + 71 |
| 5 | 84 + 81 |
| 6 | 121 + 10 |
| 7 | 164 + 36 |
| 8 | 192 + 58 |
| 9 | 33 − 30 |
| 10 | 75 − 30 |
| 11 | 75 − 5 − 40 |
| 12 | 78 − 8 − 20 |
| 13 | 78 − 24 |
| 14 | 64 − 27 |
| 15 | 120 − 29 |

**List B**

| 1 | 36 + 16 |
|---|---|
| 2 | 47 + 27 |
| 3 | 43 + 46 |
| 4 | 42 + 88 |
| 5 | 72 + 73 |
| 6 | 132 + 20 |
| 7 | 173 + 27 |
| 8 | 193 + 57 |
| 9 | 45 − 20 |
| 10 | 85 − 40 |
| 11 | 85 − 5 − 30 |
| 12 | 66 − 6 − 40 |
| 13 | 66 − 44 |
| 14 | 83 − 45 |
| 15 | 130 − 38 |

**List C**

| 1 | 49 + 29 |
|---|---|
| 2 | 38 + 18 |
| 3 | 53 + 56 |
| 4 | 37 + 83 |
| 5 | 61 + 64 |
| 6 | 124 + 10 |
| 7 | 154 + 46 |
| 8 | 183 + 67 |
| 9 | 79 − 40 |
| 10 | 95 − 50 |
| 11 | 95 − 5 − 40 |
| 12 | 79 − 9 − 50 |
| 13 | 79 − 57 |
| 14 | 93 − 67 |
| 15 | 140 − 43 |

**List D**

| 1 | 38 + 48 |
|---|---|
| 2 | 19 + 29 |
| 3 | 42 + 47 |
| 4 | 66 + 34 |
| 5 | 53 + 52 |
| 6 | 113 + 10 |
| 7 | 182 + 18 |
| 8 | 174 + 76 |
| 9 | 83 − 50 |
| 10 | 75 − 40 |
| 11 | 75 − 5 − 30 |
| 12 | 87 − 7 − 50 |
| 13 | 87 − 54 |
| 14 | 84 − 29 |
| 15 | 150 − 56 |

**EXTRA!**
Make up some sums like those in the lists of your own. Each make up two lists of 15. Try to keep them as similar to those in the lists as possible. Look carefully at each sum to see the way it works.

Now try your own lists with your partner. Record your answers on the resource sheets as before.

# Quick recall, 4

## What you need
Two copies of the *Quick recall resource sheet* (page 48), a calculator for checking.

**Tip:** Be very careful to listen to each question and do not guess it will be one type of sum or another. Plan ahead how you are going to read the sums with missing numbers in them. Make sure you both understand the way you say these sums.

You will need to work with a friend.

One of you is the reader and the other answers the questions. The reader says the first sum in List A. The partner writes the answer in the first column of the resource sheet as quickly as possible. As soon as the first question has been answered move on to the second and so on until all the questions in List A have been answered.

Then swap jobs and repeat, using List B

When you have finished both lists, swap resource sheets and check each others' answers using a calculator.

You can have another go using Lists C and D.

| | List A |
|---|---|
| 1 | $20 \div 5$ |
| 2 | $10 \div 2$ |
| 3 | $9 \div 3$ |
| 4 | $15 \div 3$ |
| 5 | $20 \div 5$ |
| 6 | $24 \div 4$ |
| 7 | $27 \div 3$ |
| 8 | $32 \div 4$ |
| 9 | $72 \div 8$ |
| 10 | $45 \div 5$ |
| 11 | $36 \div 6$ |
| 12 | $54 \div 6$ |
| 13 | $81 \div 9$ |
| 14 | $56 \div 8$ |
| 15 | $36 \div 3 \div 4$ |

| | List B |
|---|---|
| 1 | $12 \div 4$ |
| 2 | $6 \div 3$ |
| 3 | $16 \div 4$ |
| 4 | $10 \div 5$ |
| 5 | $40 \div 5$ |
| 6 | $36 \div 4$ |
| 7 | $24 \div 3$ |
| 8 | $21 \div 3$ |
| 9 | $63 \div 9$ |
| 10 | $35 \div 5$ |
| 11 | $48 \div 6$ |
| 12 | $42 \div 7$ |
| 13 | $64 \div 8$ |
| 14 | $48 \div 8$ |
| 15 | $45 \div 3 \div 5$ |

| | List C |
|---|---|
| 1 | $5 \times ? = 30$ |
| 2 | $3 \times ? = 21$ |
| 3 | $? \times 8 = 16$ |
| 4 | $? \times 9 = 36$ |
| 5 | $72 \div ? = 8$ |
| 6 | $45 \div ? = 9$ |
| 7 | $36 \div ? = 6$ |
| 8 | $36 - 17 + 23$ |
| 9 | $61 - 43 + 34$ |
| 10 | $38 + 22 - 15$ |
| 11 | $29 + 31 - 24$ |
| 12 | $53 + 62 - 19$ |
| 13 | $67 + 54 - 27$ |
| 14 | $123 - 45 + 6$ |
| 15 | $234 - 162 + 9$ |

| | List D |
|---|---|
| 1 | $4 \times ? = 32$ |
| 2 | $4 \times ? = 24$ |
| 3 | $? \times 7 = 14$ |
| 4 | $? \times 9 = 27$ |
| 5 | $63 \div ? = 9$ |
| 6 | $35 \div ? = 5$ |
| 7 | $48 \div ? = 6$ |
| 8 | $43 - 18 + 25$ |
| 9 | $72 - 34 + 45$ |
| 10 | $46 + 44 - 35$ |
| 11 | $38 + 42 - 21$ |
| 12 | $57 + 82 - 17$ |
| 13 | $74 + 68 - 32$ |
| 14 | $132 - 56 + 9$ |
| 15 | $213 - 171 + 8$ |

## EXTRA!
Make up some sums like those in the lists of your own. Each make up two lists of 15. Try to keep them as similar to those in the lists as possible. Look carefully at each sum to see the way it works.

Now try your own lists with your partner. Record your answers on the resource sheets as before.

How to be Brilliant at Mental Arithmetic

# How does it work?

Mystify your friends by always knowing the number they thought of.

Ask your friend to follow the steps carefully.

| Think of a number |
| --- |
| Add 10 |
| Multiply by 2 |
| Subtract 5 |
| Add 3 |
| Add the number you first thought of |
| Divide by 3 |
| Tell me your answer |

You can always work out their first number quickly by subtracting 6.

They will be amazed!

Here's another magical calculation.

| Think of a number |
| --- |
| Add 3 |
| Multiply by 4 |
| Subtract 8 |
| Divide by 4 |
| Tell me your answer |

The number your friend thought of is always 1 less than the answer.

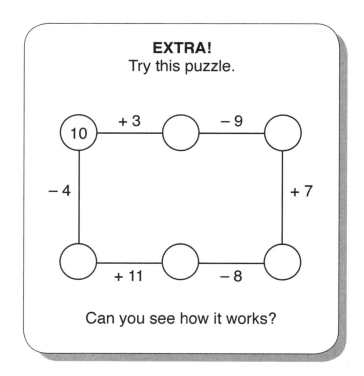

**EXTRA!**
Try this puzzle.

Can you see how it works?

# Pairs and trios, 1

### What you need
A copy of the *Pairs and trios resource sheet* (page 46).

**Tip:** A sector is a part of a circle and looks rather like a slice of pie.

Look at the circles.

See how the numbers are arranged.

Add together the pairs of numbers in each sector, eg 1 + 11. Write the answers round the outside.

What do you notice about the answers?

Now try these.

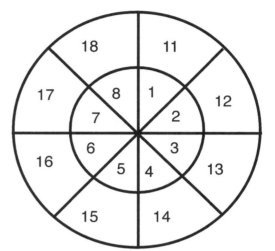

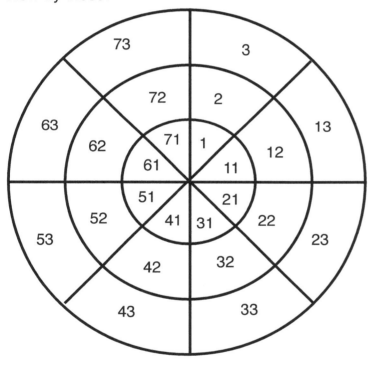

What do you notice?

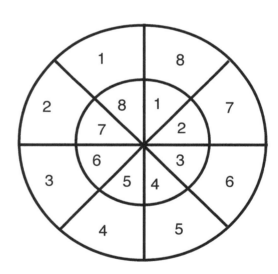

What do you notice?

---

### EXTRA!
Make up some pairs and trios using only even numbers.
Write down what you notice.

---

How to be Brilliant at Mental Arithmetic

# Pairs and trios, 2

## What you need
A copy of the *Pairs and trios resource sheet* (page 46).

Look at the circles.

See how the numbers are arranged. Multiply together the pairs of numbers in each sector of the circle, eg 1 x 8, 2 x 7.

Write the answers round the circle.

What do you notice about the answers?

Now try these:

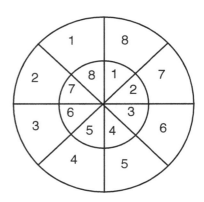

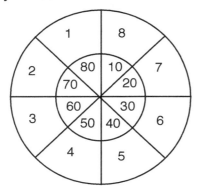

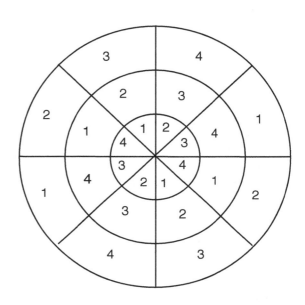

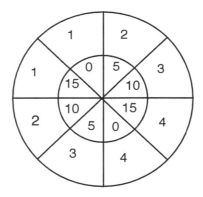

What do you notice?

Explore some pairs and trios of your own. Use patterns for the numbers in the circles and sectors.

---

**EXTRA!**
Make up some pairs and trios using only multiples of 10.
Write down what you notice.

---

# Counting on

**What you need**
A number line.

One way of quickly adding or subtracting in your head is to use the technique of counting on.

**Adding**. When adding two two-digit numbers together you can use the number line to help you add the units and then add the tens or the other way round. For instance, 56 + 27 = *83*.

Find 56 on the number line.

50    56    60

Both ways are as good as each other. You can use either.

Count on 7 to reach 63.

56    60    63

or

Count on 20 to reach 76.

56  60  66  70  76

Count on 20 to reach 83.

63    70  73    80  83

Count on 7 to reach 83.

76    80    83

Use the counting on method (either with a number line, or even better, in your head) to help you find the answers to these sums.

84 + 129          29 + 138          68 + 66
72 + 25          35 + 49          57 + 28 + 91

**Subtraction**. With subtraction there are even more ways you can use the number line. One is to count back in a similar way to the way you did the addition above. Another way is known as the 'shopkeeper's method' when you count on from the smaller number to the larger number. For instance: 100 – 34 = *66*.

Find 34 on the number line:

30    34    40

Count on until you reach a multiple of 10 (count on 6).

30    34    40

Count on until you reach 100 (count on 60).

40  50  60  70  80  90  100

or

Count on in tens until you get as close as possible to 100 (count on 60).

30 34  40  50  60  70  80  90 94  100

Count on until you reach 100 (count on 6).

90    94    100

Use this method to help you find the answers to these sums:

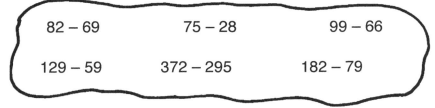

82 – 69          75 – 28          99 – 66
129 – 59          372 – 295          182 – 79

**EXTRA!**
Find out how many of your friends use the counting on method when doing sums in their heads.

How to be Brilliant at Mental Arithmetic

# Addition 100 square

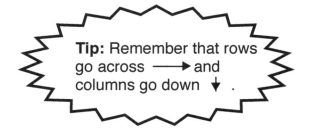

**Tip:** Remember that rows go across ⟶ and columns go down ↓ .

**What you need**
A copy of the *100 square resource sheet* (page 41), plastic counters, a calculator.

Look at the 100 square. Notice how the units digits increase as you go across the rows and the tens digits increase as you go down the columns.

Choose a starting number. It can be any number towards the middle of the 100 square. Write down your number, then cover it with a plastic counter.

Now add 34 in stages to your number. Record this as + 34. First of all, add 4 by counting four squares along the row. Now add 30 by counting down three columns. Where do you end up? Write the answer. Check it with a calculator.

For example:  18  +  34  =  52
18  ⟶  22  ↓  52

| 11 | 12 | 13 | 14 | 15 | 16 | 17 | 18 start | 19 * | 20 * |
|----|----|----|----|----|----|----|----|----|----|
| 21 * | 22 * | 23 | 24 | 25 | 26 | 27 | 28 | 29 | 30 |
| 31 | 32 * | 33 | 34 | 35 | 36 | 37 | 38 | 39 | 40 |
| 41 | 42 * | 43 | 44 | 45 | 46 | 47 | 48 | 49 | 50 |
| 51 | 52 * | 53 | 54 | 55 | 56 | 57 | 58 | 59 | 60 |

Try some more addition sums in this way. Record the answers below.

**EXTRA!**
Work out a similar method for subtracting. Explain how it works to a friend.

Explain this method of adding to a friend by speaking, drawing, demonstrating or writing.

# Splitting up sums

**What you need**
Base ten apparatus (optional), calculator.

Sums can be split up into smaller parts to make them easier. Making the sum into smaller and easier sums helps you to be able to do it more easily in your head. For instance:

39 + 12 can be thought of as (30 + 10) and (9 + 2). Each smaller sum can then be added to find the final answer.

In other words, 39 + 12 becomes 40 + 11, which makes 51 in total.

This also works with subtraction sums sometimes. For instance:

39 – 12 can be thought of as (30 – 10) and (9 – 2). These two are then added to find the final answer.

In other words, 39 – 12 becomes 20 + 7 which makes 27.

Try breaking these sums down into smaller parts to help you find the answers. You may find base 10 apparatus helpful.

**Tip:** Be careful with subtraction sums as you may find that the units calculation gives a negative answer. This makes using this method very difficult. For example: 42 – 18 would break down to (40 – 10) and (2 – 8). If this happens, it may be best to use another method.

Check your answers using a calculator.

| 27 + 38 | 83 – 41 |
|---|---|
| (   +   ) and (   +   ) makes (   +   ) | (   –   ) and (   –   ) makes (   +   ) |
| 63 + 45 | 75 – 14 |
| (   +   ) and (   +   ) makes (   +   ) | (   –   ) and (   –   ) makes (   +   ) |
| 94 + 34 | 98 – 32 |
| 127 + 134 | 179 – 32 |
| 259 + 376 | 268 – 146 |

**EXTRA!**
Try adding three numbers in this way. Are there any other ways you could think of to make adding and subtracting sums more easy to do in your head?

# Factors can help!

**Background**

In 3 x 4 = 12, the 3 and 4 are called factors of 12. This means that 12 can be divided by 3 and 4. Other factors of 12 are:       6 and 2  because  6 x 2 = 12
                                                                      12 and 1  because  12 x 1 = 12.

When doing multiplication with two digit numbers, it can sometimes help to split one of the numbers up into its factors. This will make smaller multiplication sums for you to do.

For instance:

If you are working out 37 x 18 it is useful to know that 18 = 6 x 3. You can then work out 37 x 6 first and then multiply the answer by 3, ie 222 x 3 which gives 666.

Use this method to work out the following sums.

| Sum | Factors of one number | Interim stage | Answer |
|-----|------------------------|---------------|--------|
| 37 x 18 | 6 x 3 = 18 | 37 x 6 = 222 | 222 x 3 = 666 |
| 34 x 12 | 3 x 4 = 12 | 34 x 3 = | x 4 = 408 |
| 45 x 21 | 7 x 3 = 21 | 45 x 7 = | x 3 = 945 |
| 63 x 15 |  |  |  |
| 43 x 14 |  |  |  |
| 19 x 27 |  |  |  |
| 36 x 13 |  |  |  |
| 42 x 17 |  |  |  |
| 32 x 19 |  |  |  |
| 24 x 18 |  |  |  |

**Tip:** Remember that you can do multiplication sums in any order. For instance, 25 x 3 = 75 and 3 x 25 = 75.

**EXTRA!**
Some numbers have lots of factors. Have a class competition to find the number that has the largest amount of factors.

# More and less

**What you need**
A set of number cards numbered 21-100 made from two copies
of the *Blank digit cards resource sheet* (page 45), a calculator.

This is a game to play with a friend.

Mix up the cards and lay them face down on the table. Decide what rule the game will have.
For example, 'add 10', 'take away 15', 'add 25'.

One of you turns over the first card and says the number. The other player has to apply the rule
to the number. If the first player thinks the answer is right, the second player gets to keep the
card.

If the first player thinks the answer is wrong he or she must say if the right answer is higher or
lower. This is then checked using a calculator and the player who is right gets to keep the card.
If neither player is right, the card should be returned to the bottom of the pile. Continue taking it
in turns to turn over cards. Keep the rule the same until you have used all the cards. The player
with the larger pile of cards at the end is the winner.

For example:

| The rule is 'add 22' |
| --- |

| Player A | Player B |
| --- | --- |
| Turns over 37 | |
| | Says it is 59 |
| Agrees this is right | |
| | Keeps the card |
| | |
| | Turns over 49 |
| Says it is 61 | |
| | Says this is wrong and the right answer is higher |
| Checks with a calculator and the answer is 71 | |
| | Keeps the card |

**EXTRA!**
Play the game with
only multiplication rules
or have two rules. For
example,
'add 24, then subtract
15' or 'add 7, then
multiply by 2'.

How to be Brilliant at Mental Arithmetic

# How many can I do?

**What you need**
Red and green felt-tip pens or coloured pencils.

Look at the sums below. Put a red circle around all those you 'just know the answer to'. Put a green circle around all those you can work out in your head.

| | | | | |
|---|---|---|---|---|
| 49 − 32 = | 3 x 3 = | 36 + 13 = | 9 x 6 = | 24 ÷ 3 = |
| 4 x 8 = | 65 − 44 = | 8 x 7 = | 527 + 9 = | 9 x 9 = |
| 100 ÷ 5 = | 222 + 111 = | 35 x 10 = | 63 + 31 = | 58 − 22 = |
| 87 + 53 = | 188 + 12 = | 76 − 51 = | 6 x 4 = | 58 − 22 = |
| 46 − 29 = | 10 x 10 = | 10 x 5 = | 204 + 111 = | 203 + 407 = |
| 11 x 11 = | 27 − 15 = | 37 + 19 = | 413 + 217 = | 9 x 7 = |
| 79 + 12 = | 1000 ÷ 100 = | 210 + 740 = | 35 ÷ 5 = | 555 ÷ 5 = |
| 36 ÷ 6 = | 49 − 32 = | 75 ÷ 5 = | 11856 ÷ 1 = | 68 + 43 = |
| 124 − 9 = | 874 + 26 = | 36 ÷ 9 = | 25 − 10 = | 399 + 1 = |
| 388 − 144 = | 8 x 8 = | 30 + 50 = | 7 x 7 = | 603 − 457 = |
| 15 x 10 = | 9 x 6 = | 64 ÷ 8 = | 413 − 211 = | 7 x 8 = |
| 12 − 12 = | 56 + 14 = | 6 x 6 = | 527 − 299 = | 12 ÷ 12 = |
| 5 x 7 = | 299 + 101 = | 4 + 5 + 6 = | 132 + 316 = | 53 − 28 = |
| 333 − 29 = | 508 − 211 = | 82 + 21 = | 4 x 7 = | 77 + 36 = |
| 152 + 207 = | 60 ÷ 10 = | 84 − 57 = | 20 x 5 = | 33 ÷ 3 = |

## EXTRA!
Make up a set of sums for which the answer is always a number with a 9 in it.
For instance, make up sums whose answers are: 9, 29, 391, 798, 349.
Ask you friend to see how many he or she can do in his or her head.

# It's close enough!

In everyday life we do not always need to know the exact answer to a sum. Often it is good enough to have a rough idea of the answer.

It is important to know when you need an exact answer and when an approximate answer is close enough. Finding an approximate answer is usually easier and quicker in your head than working out an exact answer on paper or with a calculator.

When working out an approximate answer you need to round one or more of the numbers in the sum up or down. For example: 155 – 38 becomes much easier to do if you round the 38 up to 40 and then do the subtraction sum.

Use rounding to help you work out these calculations in your head.

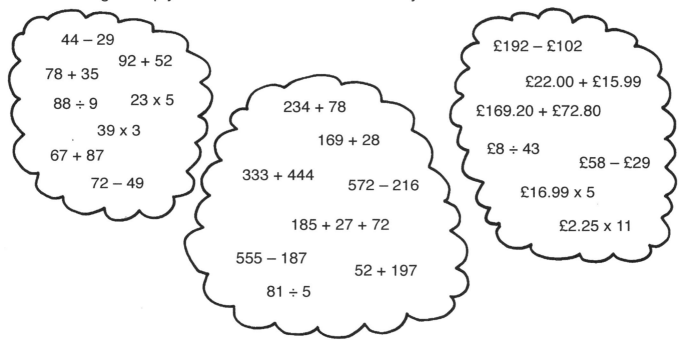

44 – 29
92 + 52
78 + 35
88 ÷ 9     23 x 5
39 x 3
67 + 87
72 – 49

234 + 78
169 + 28
333 + 444     572 – 216
185 + 27 + 72
555 – 187     52 + 197
81 ÷ 5

£192 – £102
£22.00 + £15.99
£169.20 + £72.80
£8 ÷ 43
£58 – £29
£16.99 x 5
£2.25 x 11

Use the calculator to work out the exact answers and compare with your approximate answers. How close were you?

Were your approximate answers over or under estimates?

---

**EXTRA!**
Think about times when arithmetic is used in everyday life. In how many circumstances do people need exact answers and how often will an approximate answer be good enough?
Are there times when it is important to over-estimate rather than under-estimate?

---

How to be Brilliant at Mental Arithmetic

# Butterfly pairs

Look at the butterfly. Find a pair of numbers that total 100. Colour them the same colour. Choose another colour and find another pair of numbers that add to 100. You will have to use your colours more than once.

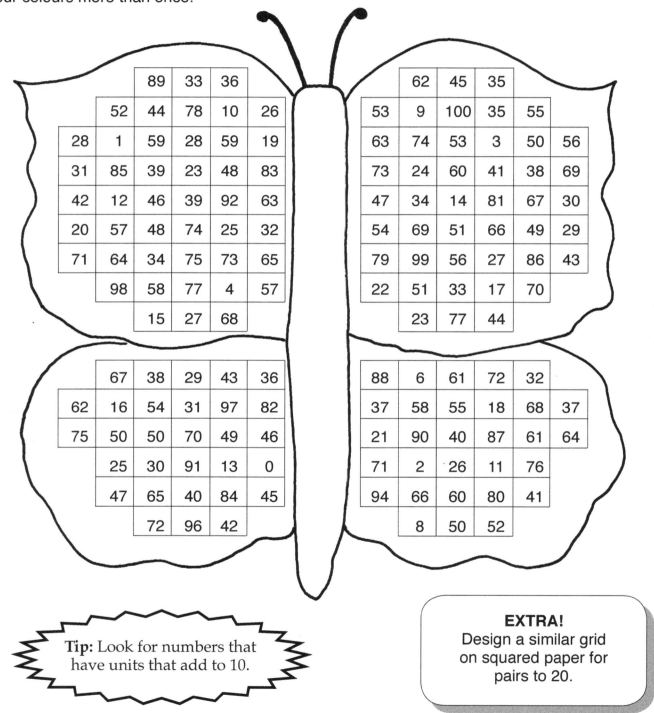

Upper left wing grid:

| | 89 | 33 | 36 | | |
| 52 | 44 | 78 | 10 | 26 | |
| 28 | 1 | 59 | 28 | 59 | 19 |
| 31 | 85 | 39 | 23 | 48 | 83 |
| 42 | 12 | 46 | 39 | 92 | 63 |
| 20 | 57 | 48 | 74 | 25 | 32 |
| 71 | 64 | 34 | 75 | 73 | 65 |
| | 98 | 58 | 77 | 4 | 57 |
| | | 15 | 27 | 68 | |

Upper right wing grid:

| | 62 | 45 | 35 | | |
| 53 | 9 | 100 | 35 | 55 | |
| 63 | 74 | 53 | 3 | 50 | 56 |
| 73 | 24 | 60 | 41 | 38 | 69 |
| 47 | 34 | 14 | 81 | 67 | 30 |
| 54 | 69 | 51 | 66 | 49 | 29 |
| 79 | 99 | 56 | 27 | 86 | 43 |
| 22 | 51 | 33 | 17 | 70 | |
| | 23 | 77 | 44 | | |

Lower left wing grid:

| | 67 | 38 | 29 | 43 | 36 |
| 62 | 16 | 54 | 31 | 97 | 82 |
| 75 | 50 | 50 | 70 | 49 | 46 |
| | 25 | 30 | 91 | 13 | 0 |
| | 47 | 65 | 40 | 84 | 45 |
| | | 72 | 96 | 42 | |

Lower right wing grid:

| 88 | 6 | 61 | 72 | 32 | |
| 37 | 58 | 55 | 18 | 68 | 37 |
| 21 | 90 | 40 | 87 | 61 | 64 |
| 71 | 2 | 26 | 11 | 76 | |
| 94 | 66 | 60 | 80 | 41 | |
| | 8 | 50 | 52 | | |

**Tip:** Look for numbers that have units that add to 10.

# Wall

**What you need**
Copies of the *Wall resource sheet* (page 47).

Look at the wall below. What numbers have been used?

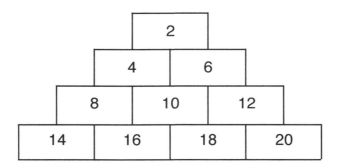

Find all the multiples of 4. Can you predict the next row? Try it out using the resource sheet. Keep going until you have completed the blank wall.

Now try this one.

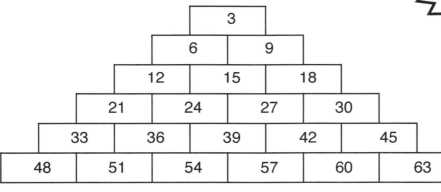

**Tip:** Some multiples of 3 are 3, 6, 9, 12, etc. Some multiples of 5 are 5, 10, 15, 20.

Look for patterns of multiples of numbers.

Continue the wall on a blank copy of the resource sheet.

**EXTRA!**
Try using other patterns and shapes.

How to be Brilliant at Mental Arithmetic

# Sports survey

The children in Years 5 and 6 of Fitwell School held a survey to see which sports clubs the children would prefer to be held at the school. The four possible clubs were table tennis, netball, football and swimming. This is what they found.

**Ash class**

|  | All | Boys | Girls |
|---|---|---|---|
| Table tennis |  | 2 | 2 |
| Netball |  | 0 | 3 |
| Football |  | 8 | 2 |
| Swimming |  | 5 | 6 |

**Beech class**

|  | All | Boys | Girls |
|---|---|---|---|
| Table tennis |  | 3 | 0 |
| Netball |  | 0 | 6 |
| Football |  | 4 | 2 |
| Swimming |  | 7 | 7 |

**Cedar class**

|  | All | Boys | Girls |
|---|---|---|---|
| Table tennis |  | 0 | 2 |
| Netball |  | 0 | 2 |
| Football |  | 6 | 2 |
| Swimming |  | 6 | 8 |

**Damson Class**

|  | All | Boys | Girls |
|---|---|---|---|
| Table tennis |  | 3 | 2 |
| Netball |  | 0 | 3 |
| Football |  | 11 | 2 |
| Swimming |  | 4 | 5 |

Complete the tables to find out how many boys and girls altogether chose each club.
Try to answer these questions about the survey as quickly as you can. Do the calculations in your head.

1    How many children were there altogether in these four classes?

2    Which club was the most popular with all the children?

3    How many boys wanted a football club?

4    How many girls wanted a football club?

5    How many more boys than girls wanted a football club?

6    Which was the least favourite club?

7    In which classes did more girls than boys want a swimming club?

8    How many children in Damson class would attend the table tennis club?

9    Which club would have the fewest boys?

10   Which club was the most popular with the girls?

---

**EXTRA!**
Make up some more questions of your own.

---

# Square numbers

**Remember:** A square number is made up of a number multiplied by itself, eg 25 = 5 x 5, 16 = 4 x 4. It can always be drawn as a square.

3 x 3    9

6 x 6    36

| 1 | | 3 | | 5 | | 7 | | 9 | | 11 | | 13 | | 15 | | 17 | | 19 |
|---|---|---|---|---|---|---|---|---|---|----|---|----|---|----|---|----|---|----|

**Add together:**

the first two odd numbers          1 + 3 = 4

the first three odd numbers        1 + 3 + 5 = 9

**Tip:** Add the next odd number of the previous answer.

the first four odd numbers         1 + 3 + 5 + 7 =

the first five odd numbers         1 + 3 + 5 + 7 + 9 =

the first six odd numbers          1 + 3 + 5 + 7 + 9 + 11 =

the first seven odd numbers        1 + 3 + 5 + 7 + 9 + 11 + 13 =

the first eight odd numbers        1 + 3 + 5 + 7 + 9 + 11 + 13 + 15 =

the first nine odd numbers         1 + 3 + 5 + 7 + 9 + 11 + 13 + 15 + 17 =

the first ten odd numbers          1 + 3 + 5 + 7 + 9 + 11 + 13 + 15 + 17 + 19 =

What do you notice?

Check that the answers are square numbers and write the multiplication near the odd number series. For example:

1 + 3 = 4          (2 x 2)
1 + 3 + 5 = 9      (3 x 3)

---

**EXTRA!**
Use the 10 x 10 square on the *100 Square resource sheet* (page 41)
to investigate why adding odd numbers gives a square number.

---

How to be Brilliant at Mental Arithmetic

# Adding columns of numbers

A quick way to add a long column of numbers is to look for pairs of digits that make 10. Cross them off and make a tally mark beside the column. That way you will know exactly how many pairs that make ten you have found.

For instance:

**Tally**

\ \ \

Total of units digits
30 + 6 = 36

Use a similar method for the tens column. Look for pairs that make 100.

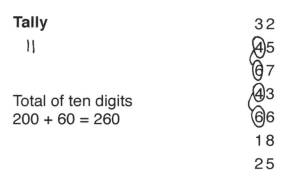

**Tally**

\|

Total of ten digits
200 + 60 = 260

```
3 2
4 5
6 7
4 3
6 6
1 8
2 5
```

Use this method to add the following restaurant bills.

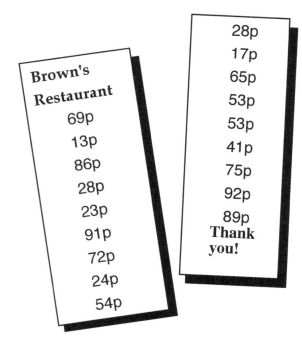

**Brown's Restaurant**

69p
13p
86p
28p
23p
91p
72p
24p
54p

28p
17p
65p
53p
53p
41p
75p
92p
89p
**Thank you!**

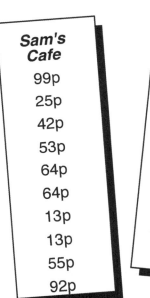

*Sam's Cafe*

99p
25p
42p
53p
64p
64p
13p
13p
55p
92p

87p
75p
56p
64p
51p
29p
65p
33p
58p
43p
Please come again!

**EXTRA!**
Find a long till receipt from a supermarket. Cover up the final total and use this method to see how quickly you can find the total yourself.

# 24 hour clock

To convert from the 12 hour clock to the 24 hour clock add 12 to the hours.

For instance: 9 o'clock converts to 21.00 (9 + 12 = 21).

To convert from the 24 hour clock to the 12 hour clock, subtract 12 from the hours.

For instance: 19.30 converts to half past 7 (or 7.30 pm) (19 − 12 = 7).

Convert these times:

**Tip:** Remember that 9 am is 09.00 and 9 pm is 21.00.

Remember that the 12 and 24 hour clocks use different ways of recording time such as o'clock, half past, quarter to.

Also, 09.00 can be read as O nine hundred hours.

8.30 am

10.15 am

11.35 am

12.05 pm

14.50

18.10

20.30

21.45

---

**EXTRA!**
Investigate how times are recorded and spoken in France. French people do not always use the 24 hour clock.

How to be Brilliant at Mental Arithmetic

# Local trains

The timetable for the trains between Gillingham and London is below.

| Gillingham | 0727 | – | 0747 | – | 0830 |
|---|---|---|---|---|---|
| Chatham | 0731 | – | 0753 | – | 0834 |
| Rochester | – | 0735 | 0755 | 0815 | 0836 |
| Sole Street | – | 0747 | 0806 | 0827 | 0847 |
| Meopham | 0744 | 0749 | 0808 | 0829 | 0849 |
| Longfield | 0749 | 0753 | 0812 | – | 0853 |
| Farningham Road | – | 0757 | 0816 | 0837 | 0857 |
| Swanley | 0757 | 0802 | 0821 | 0842 | 0902 |
| Bromley South | 0808 | 0815 | 0833 | 0854 | 0914 |
| London Victoria | 0831 | – | 0857 | 0915 | 0938 |

Read the timetable to answer these questions.

1    If you catch the 0727 from Gillingham, what time do you arrive at Swanley?

2    How long does the 0747 from Gillingham take to get to London Victoria?

3    Does it always take the same amount of time to get from Gillingham to London Victoria?

4    You are meeting a friend at Bromley South station at half past eight. What time will you leave Longfield?

5    You arrive five minutes late for your train from Chatham. You should have arrived in London at 0857. You catch the next train from Chatham. How late will you arrive?

6    How long does the 0815 from Rochester take to get to Farningham Road?

7    You miss the 0812 from Longfield by one minute. How long will you wait for the next train?

---

**EXTRA!**
Investigate how long it takes to get by train from central London
to the station nearest your school.

---

# Metric and imperial weights

## Rule
To convert metric kilograms to imperial pounds you should multiply by 11 and then divide by 5.

**Tip**: Round 13.2 to 13.0.

For instance:

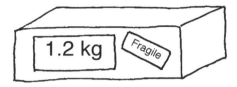
1.2 kg — Fragile

1.2 x 11 = 13.2
13.2 ÷ 5 = about 2.6 pounds

Convert these weights using the same method.

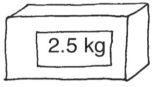
2.5 kg

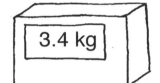
3.4 kg

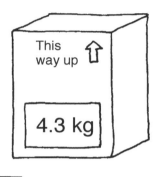
This way up — 4.3 kg

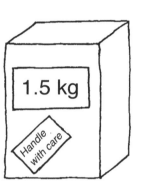
1.5 kg — Handle with care

## Rule
To convert imperial pounds to metric kilograms you should multiply by 5 and then divide by 11.

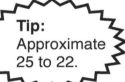

**Tip:** Approximate 25 to 22.

For instance:

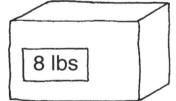
5 lbs
*Urgent Delivery*

5 x 5 = 25
25 ÷ 11 = about 2 kilograms

Convert these weights using the same method.

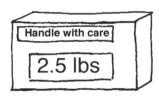
Handle with care
2.5 lbs

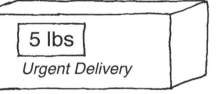
8 lbs

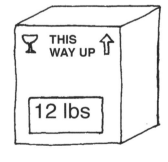
THIS WAY UP — 12 lbs

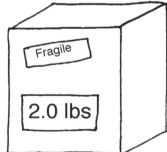
Fragile — 2.0 lbs

## EXTRA!
Weigh some items in the classroom in kilograms.
Convert the weights to imperial pounds.

How to be Brilliant at Mental Arithmetic

# Metric and imperial distance

**Rule**
To convert metric kilometres to imperial miles you should multiply by 3 and then divide by 5.

For instance:

> 20 km

$20 \times 3 = 60$
$60 \div 5 = 12$ miles

Convert these distances using the same method.

> 35 km

> 15 km

> 29 km

> 8.5 km

**Rule**
To convert imperial miles to metric kilometres you should multiply by 5 and then divide by 3.

**Tip:**
Approximate
50 to 48.

For instance:

> 10 miles

$10 \times 5 = 50$
$50 \div 3 =$ about 16 kilometres

Convert these distances using the same method.

> 22 miles

> 17 miles

> 30 miles

> 8 miles

**EXTRA!**
Look up some distances in a road atlas.
Convert the distances to metric kilometres.

# Metric and imperial capacity

**Rule**
To convert metric litres to imperial pints you should multiply by 7 and then divide by 4.

For instance:

5 x 7 = 35
35 ÷ 4 = about 9 pints

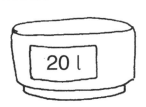
**Tip**: Approximate 35 to 36.

Convert these capacities using the same method.

 15 l

 10 l

20 l

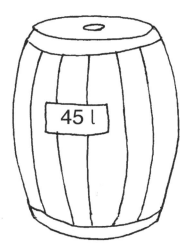 45 l

**Rule**
To convert imperial pints to metric litres you should multiply by 4 and then divide by 7.

For instance:

 16 pints

16 x 4 = 64
64 ÷ 7 = about 9 litres

**Tip:** Approximate 64 to 63.

Convert these capacities using the same method.

 5 pints

 1 pint

 100 pints

3 pints

**EXTRA!**
Measure some capacities in the classroom in litres.
Convert the capacities to imperial pints.

How to be Brilliant at Mental Arithmetic

# Converting money, 1

**Tourist conversion rate**
Iceland  £1 = 182.81 króna (ISK)

August 2010

Exact conversions for foreign currencies are usually too difficult to work out in your head quickly. People tend to use rounded amounts which make the calculations less complex so they can do them in their heads. This gives a less accurate answer, but is usually good enough for most people. For instance, the conversion rate for Icelandic Krona to British pounds rounds fairly accurately to 200 ISK to 1 pound.

Use the rounded amount to convert these prices in króna to pounds and pence.

| Item | ISK | £ | Item | ISK | £ |
|---|---|---|---|---|---|
| Peaches | 280 | | Cola | 250 | |
| Bread | 35 | | Butter | 180 | |
| Jam | 315 | | Ice-cream | 565 | |
| Crisps | 150 | | Ketchup | 275 | |
| Corn flakes | 375 | | Water | 49 | |
| Eggs | 190 | | Potatoes | 60 | |

**EXTRA!**
Investigate the prices in British shops for these items and compare the price in this country to the price converted from krona.

# Converting money, 2

On holiday in different countries five friends bought some gifts for each other. Using the currency conversion table, work out about how much each friend spent in pounds sterling (British money). Round the prices up or down to make it easier to do the calculations in your head.

| Country | 1 pound sterling can be exchanged for: |
|---------|----------------------------------------|
| USA | $1.54 (dollars) |
| Denmark | 8.70 kr (krone) |
| Lebanon | 2,318 L.L (Lebaneses pounds) |
| Iceland | 182.81 ISK (króna) |

August 2010

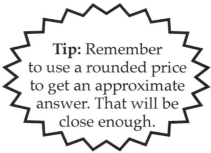

**Tip:** Remember to use a rounded price to get an approximate answer. That will be close enough.

| Iceland | Denmark | Lebanon | USA | England |
|---------|---------|---------|-----|---------|
| Sweets 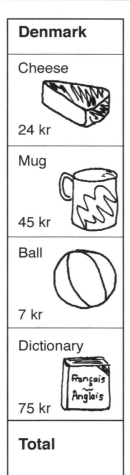 250 ISK | Cheese 24 kr | Poster  4,500 L.L | Cuddly toy 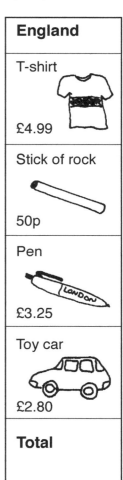 $9.50 | T-shirt £4.99 |
| T-shirt 1,200 ISK | Mug 45 kr | Sweets 1,500 L.L | CD $7.00 | Stick of rock 50p |
| Zither 3600 ISK | Ball 7 kr | Book 14,000 L.L | Sun-glasses $4.00 | Pen £3.25 |
| Icelandic doll 890 ISK | Dictionary  75 kr | Game 8,200 L.L | Baseball cap $6.90 | Toy car £2.80 |
| **Total** | **Total** | **Total** | **Total** | **Total** |

---

**EXTRA!**

Currency conversion rates for different countries are displayed in banks and printed in many newspapers. Investigate if they change from day to day.

How to be Brilliant at Mental Arithmetic

# Self-assessment sheet

Name: _____ Date: _____

| I can… | Date |
|---|---|
| recall addition bonds to 10 | |
| recall subtraction bonds to 10 | |
| recall addition bonds to 20 | |
| recall subtraction bonds to 20 | |
| recall addition bonds to 100 | |
| recall subtraction bonds to 100 | |
| quickly add together two two-digit numbers in my head | |
| quickly subtract two two-digit numbers in my head | |
| quickly pair numbers which add to 100 | |
| recall multiplication bonds to 5 x 5 | |
| recall multiplication bonds to 10 x 10 | |
| recall division facts in multiplication tables up to 5 x 5 | |
| recall division facts in multiplication tables up to 10 x 10 | |
| use rounding to help me add and subtract | |
| use rounding to help me multiply and divide | |
| use addition short-cuts to help me add numbers with units digits of 8 or 9 | |
| use addition short-cuts to help me add numbers with units digits of 6 or 7 | |
| use a 100 square to help me use counting on when adding two numbers | |
| use counting on to help me add two numbers | |
| use counting on to help me subtract two numbers | |
| use factors to help me multiply two numbers | |
| use my skill at mental arithmetic to answer questions about numerical information quickly | |
| add long columns of numbers quickly | |
| find square numbers using odd numbers | |
| convert 12 and 24 hour clocks in my head | |
| calculate lengths of time from a timetable in my head | |
| convert imperial and metric weights approximately in my head | |
| convert imperial and metric distances approximately in my head | |
| convert imperial and metric capacities approximately in my head | |
| convert foreign currency approximately in my head | |

# 100 square resource sheet

| | | | | | | | | | |
|---|---|---|---|---|---|---|---|---|---|
| 1 | 2 | 3 | 4 | 5 | 6 | 7 | 8 | 9 | 10 |
| 11 | 12 | 13 | 14 | 15 | 16 | 17 | 18 | 19 | 20 |
| 21 | 22 | 23 | 24 | 25 | 26 | 27 | 28 | 29 | 30 |
| 31 | 32 | 33 | 34 | 35 | 36 | 37 | 38 | 39 | 40 |
| 41 | 42 | 43 | 44 | 45 | 46 | 47 | 48 | 49 | 50 |
| 51 | 52 | 53 | 54 | 55 | 56 | 57 | 58 | 59 | 60 |
| 61 | 62 | 63 | 64 | 65 | 66 | 67 | 68 | 69 | 70 |
| 71 | 72 | 73 | 74 | 75 | 76 | 77 | 78 | 79 | 80 |
| 81 | 82 | 83 | 84 | 85 | 86 | 87 | 88 | 89 | 90 |
| 91 | 92 | 93 | 94 | 95 | 96 | 97 | 98 | 99 | 100 |

How to be Brilliant at Mental Arithmetic

# Digit cards 0 – 9 resource sheet

| | | | |
|---|---|---|---|
| 0 | 1 | 2 | 3 |
| 4 | 5 | 6 | 7 |
| 8 | 9 | 0 | 1 |
| 2 | 3 | 4 | 5 |
| 6 | 7 | 8 | 9 |

# Digit cards 0 – 19 resource sheet

| | | | | |
|---|---|---|---|---|
| 0 | 1 | 2 | 3 | 4 |
| 5 | 6 | 7 | 8 | 9 |
| 10 | 11 | 12 | 13 | 14 |
| 15 | 16 | 17 | 18 | 19 |
| 0 | 1 | 2 | 3 | 4 |
| 5 | 6 | 7 | 8 | 9 |
| 10 | 11 | 12 | 13 | 14 |
| 15 | 16 | 17 | 18 | 19 |

# Product square resource sheet

| x | 1 | 2 | 3 | 4 | 5 | 6 | 7 | 8 | 9 | 10 |
|----|---|---|---|---|---|---|---|---|---|----|
| 1 | | | | | | | | | | |
| 2 | | | | | | | | | | |
| 3 | | | | | | | | | | |
| 4 | | | | | | | | | | |
| 5 | | | | | | | | | | |
| 6 | | | | | | | | | | |
| 7 | | | | | | | | | | |
| 8 | | | | | | | | | | |
| 9 | | | | | | | | | | |
| 10 | | | | | | | | | | |

How to be Brilliant at Mental Arithmetic

# Blank digit cards resource sheet

How to be Brilliant at Mental Arithmetic

# Pairs and trios resource sheet

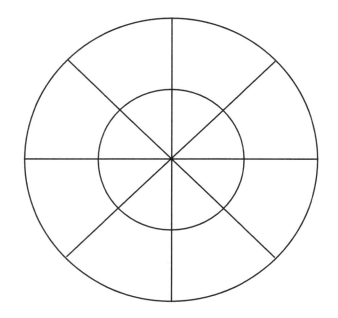
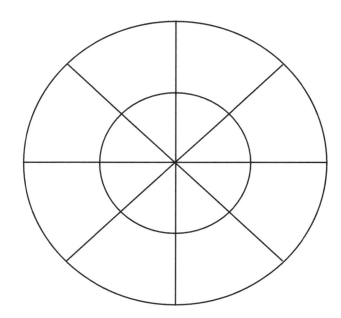

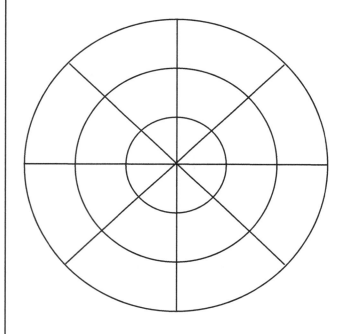
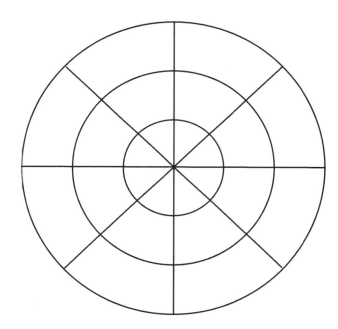

# Wall resource sheet

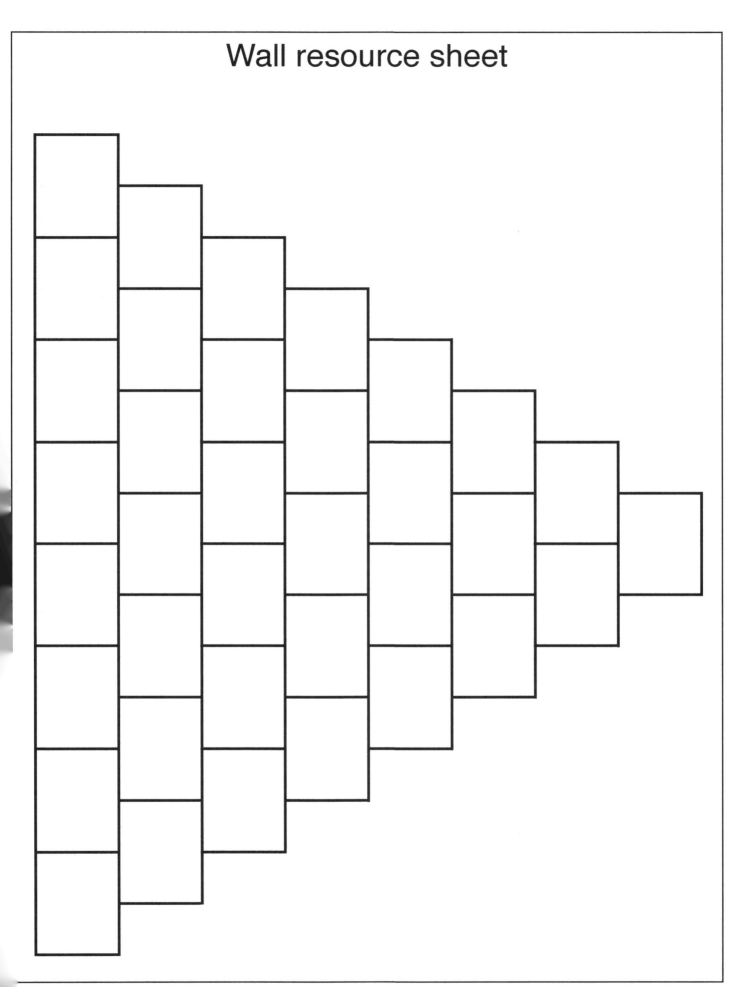

How to be Brilliant at Mental Arithmetic

# Quick recall resource sheet

Name: _____ Date: _____

| Question | 1 | 2 | 3 | 4 |
|---|---|---|---|---|
| 1 | | | | |
| 2 | | | | |
| 3 | | | | |
| 4 | | | | |
| 5 | | | | |
| 6 | | | | |
| 7 | | | | |
| 8 | | | | |
| 9 | | | | |
| 10 | | | | |
| 11 | | | | |
| 12 | | | | |
| 13 | | | | |
| 14 | | | | |
| 15 | | | | |
| Total number correct | | | | |